Oliver Thomas Osborne

A Case of Acromegaly Autopsy

Oliver Thomas Osborne

A Case of Acromegaly Autopsy

ISBN/EAN: 9783337815646

Printed in Europe, USA, Canada, Australia, Japan

Cover: Foto ©berggeist007 / pixelio.de

More available books at **www.hansebooks.com**

ACROMEGALY

THE SKELETON.

BY

OSBORNE, M.D.

A CASE OF ACROMEGALY: AUTOPSY: SKELETON.

BY O. T. OSBORNE, M.D.,

PROFESSOR OF MATERIA MEDICA AT YALE UNIVERSITY, NEW HAVEN

This case was first reported by me in *The American Journal of the Medical Sciences* for June, 1889, and since that date has been more or less constantly under my supervision.

Three years ago he had several bad attacks of epistaxis and a great deal of muscular prostration; he then for a time regained his muscular power, and was able to resume his work as a machinist, working pretty constantly during the last two years. Ever since he has been under my observation he has had short periods of passing diminished quantities of urine, and has complained of some lumbar pain. He has never been free from a constant almost unceasing diabetes aurium. He has had occasional attacks of dizziness during the last years, but there have never been any signs of mental aberration other than the temporary derangement lasting one month, in 1887, as reported by me in my first article. The last few years he has not had as much headache as in previous years.

Since the report in 1889 he has increased a great deal in size and weight.

In May, 1896, he went to the New Haven Hospital, where he remained for five days only, complaining of pain in the right ear and dozing in both ears. On May 20th a trace of albumin was found in the urine, also a trace of sugar and some granular casts. On May 23d the records show no albumin and no sugar.

During the last six months this patient has been occasionally to my office for minor conditions, such as constipation, a slight cold, etc., but on December 1, 1896, he sent for me, showing considerable œdema of the feet, legs, and scrotum, but none of the upper part of the body. Examination of the heart, as taken from my records, showed on lesion, although the sounds were rather muffled and indistinct. The pulse was good, but slow, registering 66. The amount of urine passed was scanty, but on repeated examinations during the next three weeks showed no casts and no albumin until the day before death, when 1 per cent of albumin was found, but no casts and no sugar. He did not show any signs of uræmia. A few days before his death the œdema of the face and legs became very much increased and there was also a considerable amount of acid in the fluid in the abdomen, and some slight œdema of the lungs. The patient died in December 2* 1860 suddenly and unexpectedly with signs of simple syncope; in other words, he died in the manner described by Marie in his early description of this disease. There was no previous history whatever of cardiac disease or angina.

He was forty-seven years old at the time of his death and had had his disease for twenty three years, the first symptoms having appeared at the age of twenty-four years.

The post-mortem was made thirty hours after death, and showed that he died of acromegaly and not from any single complication or concurrent disease.

His estimated weight was at least 300 pounds and the measurements of the body are much greater than those recorded in 1892.

At the level of the ensiform cartilage the body is fifty-three inches in circumference (forty-five inches in 1892). The circumference of the chest at the nipple line is fifty-one and one-half inches (forty-five and one half inches in 1892). The circumference of the neck is twenty-one inches. The circumference of the cranium around the occiput and superciliary ridges is twenty-six inches (twenty-four inches in 1892).

Rigor mortis was almost normal. The scalp was thick and covered with an abundant growth of coarse hair. The abdomen was greatly distended, and the lower extremities, scrotum and penis were very œdematous. The blood in the legs was in the arteries and not in the veins, probably on account of the œdema so pressing upon the veins to prevent the necessity of the arteries sending the blood into the veins after death. The fat of the abdominal walls was considerable in amount, and largely injected with blood, the vessels being dilated. There was a large amount of ascitic fluid in the abdominal cavity.

On cutting through the ribs the medulla of the bones squeezed out, being soft, and of a reddish color. In the right pleural cavity was a large amount of dark-colored serum, and the left pleura cavity contained a small amount of yellowish serum, as did the pericardium.

The heart was enormous, being fifteen inches in circumference at the base of the ventricles, and weighed two pounds and nine ounces. From the auriculo-ventricular junction over the left ventricle to the apex is seven and one half inches, while the length of the interventricular septum is eight and three quarters inches. The right side of the heart, both auricle and ventricle, was filled with blood. The wall of the right ventricle was three-eighths of an inch thick, and the wall of the left ventricle was seven-eighths of an inch thick. The valves of the heart were normal except some slight thickening at the edges of the aortic valves, with a small calcification half an inch from the valve on the aortic wall; otherwise the aorta is not diseased. The coronary artery is enlarged, admitting an ordinary lead pencil into the orifice. No occlusion of the heart-vessels was found.

The left lung weighed two pounds and fifteen ounces, and was normal throughout, and at the distal end a distinct amount of edema. The right lung showed a considerable consolidation at its apex posteriorly, which however was easily broken down. There was some fibrous adhesions and old caseous deposits at the apex of the lung.

The intestines were normal, except although the adhesions were found a good many irregular peritoneal adhesions (the clinical history showed frequent abdominal pain). The large intestine was adherent to the spleen and to the body of the liver.

The stomach was dilated, and the walls were thinner than normal, with no signs of acute or chronic congestion. The spleen was enlarged and dilated, and it weighed two pounds and four ounces. The capsule was not adherent, but was thickened. The splenic substance was quite hard in consistency, and cut with a smooth surface, the cut surfaces being quite dark; there was considerable connective tissue, apparently throughout.

The liver was enormously enlarged, weighing seven pounds and two ounces. The liver felt hard, and the capsule was slightly thickened; the cut surface was light in color, with a slightly mottled appearance.

The gall bladder was dilated and hung below the liver edge, and was filled with gall stones. The suprarenal capsules were of normal size.

The right kidney contained three large cysts which were evident before the capsule was removed; one of these cysts would hold a small egg. The cortical substance was thicker than normal. The weight of the right kidney was eleven ounces after the cystic fluid was removed. The left kidney weighed ten and one-half ounces and presented no evidence of the external surface, but one small cyst was found on opening it.

The thyroid gland was enlarged and weighed 101 grammes. A large gland, weighing 200 grammes, which proved to be an extra-thyroid, was found in the median line high up in the chest cavity, just above the upper end of the sternum. The left vagus nerve was enlarged and showed a marked swelling for the lack of its cranial.

Cranium. The surfaces of bone emerged over in several places had become loose. The bones of the cranium were much softer than normal, while the spongy tissue was increased in thickness and very vascular. The thickest part of the cranium was found at the right side of the head in the occipital bone.

There was a calcification in the dura mater directly over the longitudinal sinus about on a line with the external ears, the greatest amount of this calcification being at the right of the longitudinal sinus. The brain was not injected more than normal, and without the dura mater weighed three pounds.

The convolutions of the brain were normal.

The right vertebral artery showed a cylindrical enlargement for three-eighths of an inch, beginning one and one-quarter inches from its junction with the opposite vertebral artery. The coats of the artery at this part were thickened and hard. The whole right vertebral artery was distinctly larger than the left, while the left vertebral artery was quite small for an inch before its junction with the right vertebral. Also there was a distinct narrowing of the left artery about an inch before its junction with the right vertebral artery, and it was distinctly displaced from the median line toward the left, with its convexity toward the left. The two posterior cerebral arteries were of the same size, while the posterior communicating arteries, if present, were very small. The other parts of the circle of Willis were normal. The cranial nerves all appeared normal. A small tumor about the size of a pea was attached by a few fine fibres in front of the right crus cerebri at the upper part, behind the right optic tract, and lateral to the right corpus albicans.

In the right hemisphere of the cerebrum, at the upper end of the precentral sulcus, and extending more in front of this than behind, alongside the longitudinal fissure, was a roughened spot, measuring about seven-eighths of an inch in diameter, corresponding to the roughened area in the dura mater. A similar spot, somewhat smaller, was in a corresponding position on the left hemisphere. A needle thrust through the corpus callosum to the base of the brain, between the centres of these areas, comes out in the position of the pituitary body.

The superior and inferior longitudinal sinuses were normal, but the tentorium cerebelli was thickened at its attachments.

The velum interpositum and choroid plexus were thicker than normal.

Careful examination showed nothing abnormal in the substance of the cerebrum or cerebellum.

Sella turcica and its boundaries. In the sella turcica, which was very much enlarged, was a pulpy mass, containing a cyst, which was ruptured in removing the brain. This cyst contained yellow serum in [illegible].

The cavernous sinuses and internal carotids were all pressed upon by the pituitary cyst, most marked on the right side. There were several little tumors attached to the inner surface of the dura mater covering the dorsum sellae. There were also several small tumor growths on membrane covering the pituitary body. There were more fibrous bands than usual in the cavernous sinuses, especially on the right side, so that the canal was much contracted. This was true on both sides, but most decidedly on the right.

The cyst-cavity in the sella turcica was S-shaped, enlarged at both ends, dipping down nearly to the bottom of the sella turcica, and bending around the right internal carotid forward and lying between the internal carotid on the inside and the cavernous sinus on the outer side. The right Gasserian ganglion seemed pushed a little outward and forward.

I am indebted to Dr. C. J. Bartlett, Assistant Professor of Pathology at Yale, for the following report of the examination of the specimens taken at the autopsy:

THE PATHOLOGICAL FINDINGS BY DR. C. J. BARTLETT. All of the tissues received for microscopic examination were either put at once into Zenker's fluid, or into Müller's fluid for a day or two and then into Zenker's. This consists of Müller's fluid saturated with corrosive sublimate, to each 100 c.c. of which 5 c.c. of glacial acetic acid is added when used. After being in the for twenty-four hours and then washed in running water for twenty-four hours, they were hardened in alcohol and imbedded by the celloidin process. The corrosive sublimate is removed by treating the sections before staining with Gram's iodine potassium iodide solution for fifteen minutes, and then rinsing in alcohol.

The chief pathological changes are seen in the contents of the sella turcica and adjacent tissues, in the thyroid body, and in the presence of a thoracic thyroid gland.

Sections of the adenomatous taken from the sella turcica (Plate I., Fig. 1) consist chiefly of cells showing a frequently of arrangement and with almost no fibrous tissue between them, but leaving a small amount of granular intercellular material. The cells are mostly in contact with each other as a rule, but between them spaces are left, these in some areas occupying a larger part of the field in some portions of the section than the cells do. Thinwalled bloodvessels are numerous, although blood-corpuscles are not present in the unaltered vessels, nor between the cells of the tissue. In places the bloodvessels are in contact with the cells, in others they run through open spaces with apparently no other support than that of the endothelial cells forming their walls. The cells of the section vary greatly in size and appearance. In shape the greater number are either round or oval. The smallest consist chiefly of a single deeply staining round nucleus, five or six microns in diameter, surrounded by a thin border of granular protoplasm. At the other extreme in size are cells twenty microns or more in diameter, containing from four to ten small, deeply staining round or oval nuclei, the nuclei in some being crowded at one side of the cell or arranged at and the periphery, in others being scattered throughout the cell protoplasm. A majority of the cells are intermediate in size, containing from two to six nuclei and having a granular protoplasm.

A section through the Gasserian ganglion and cavernous sinus of the right side (Plate I., Fig. 2 and 3) show a mass of cells quite similar to those described above. Between the bundles of nerve fibres of the fifth nerve, which the wall of the cavernous sinus is largely replaced by this cellular growth, which is evidently an invasion of the adjacent tissue by the growth in these cells nuclei. The nuclei of the cells are many of them slightly larger than in the section above described, and are more vesicular, not staining so deeply. A few of the cells consist almost wholly of a collection of these vesicular nuclei, four or more in number, massed together with little cell-protoplasm between or around them. The thin-walled bloodvessels are filled with blood, and in places the red blood corpuscles lie between the cells or in clumps.

Sections through the left Gasserian ganglion show a smaller number of similar cells, the new growth not having invaded the surrounding tissues as widely here as on the right side. In sections of both the right and left Gasserian gang in the nerve-tissue appears normal.

The small tumors with which was associated in the right ptos cerebri (Plate I., Fig. 4), as well as those from the dura mater cellae, show a similar structure to that described, the vesicular nuclei and the masking of nevers, of here in a single cell surrounded by only a small amount of protoplasm, being even more noticeable than in transections of the Gasserian ganglia.   In places there is considerable hemorrhage into the tissue.

The source and classification of the growth filling the vesicles here, and invading the surrounding tissues, with the smaller secondary tumors elsewhere, are somewhat doubtful.   No regular grouping of the cells is noticeable aside from a slight column-ar arrangement in places, in the small vessels, how entire of some of the columns of cells being occupied by small blood-vessels; and probably it is, in the closest meaning the vessels as, but whether end otherwise tous in origin, or not, could not be determined.   The gross appearance of the small nodular growths on the dura-mater was similar to that seen in the secondary endothelioma of the lobes of the peritoneum, and produced a prejudice in favor of such a diagnosis, but this could not be confirmed microscopically.

In the thyroid gland (Plate I., Fig. 5) the most noticeable change from normal is a marked increase in the connective tissue there being comparatively thick bundles of this between the vesicles.   The larger vascular spaces contain colloid material, while at smaller are filled with groups of epithelial cells.

The gland found in the upper part of the thoracic cavity (Plate II., Fig. 6) shows a distinct division into cortical and medullary portions the latter constituting somewhat more than half the diameter of the entire body, and being made up of fibrous tissue.   The cortical portion resembles that of developing thyroid tissue as described by Halsted as being found in dogs after partial extirpation of the thyroid gland.   This gland is surrounded by a fibrous capsule, richly supplied with blood-vessels, and bundles of fibrous tissue extend from this into the cortical portion.   This cortical portion is composed of vesicular spaces, varying in size from those a mm. long to those of microscopic dimensions, between which are thin bundles of fibrous tissue with numerous small blood-vessels.   The spaces are lined with cells cuboidal or slightly columnar in shape, each cell containing a single deeply staining nucleus.   The cells may be one layer deep, or there may be several layers, and many of the smaller spaces are completely filled with cells.   The larger spaces are oval in shape elongated perpendicularly to the surface of the gland, and most of them contain more or less homogeneous colloid material, some of them being only partly filled with it.   Others are filled with cells, or partly with colloid material and partly with degenerating cells, the outlines of which may still be made out.

Of the nerve-tissues sections were made and examined by Dr. Robert L. Peck, from both Gasserian ganglia from different portions of the medulla, and the upper end of the spinal cord, from the pons, the right crus cerebri, the left crus cerebri the median nerve of the left side, and from the posterior tibial nerve of the right side. These all showed normal structure except the lower end of the medulla, the median nerve, and a small branch of the posterior tibial nerve.

A cross-section of the lower end of the medulla showed, to the naked eye, an atrophy of the posterior column of the left side for the extent of 6 mm. Microscopically there was seen to be a disappearance of nerve-fibres in this portion, and an increase of fibrous tissue, this increase not being sufficient to replace the nerve tissue that had degenerated, leaving the left posterior column smaller than the right. The left median nerve had undergone such post-mortem changes before hardening that a minute microscopic examination was impossible. The whole section, however, showed a great increase in fibrous tissue between the bundles of the nerve-fibres.

A small branch of the posterior tibial nerve showed sclerosis. The pineal gland did not show any pathological changes.

Sections through the heart-wall (left ventricle) show a marked increase in thickness of both the pericardium and endocardium and an increase in fibrous connective tissue between the muscle fibres, most marked near the endocardium. In some of these fibrous bands are collections of fat-globules. The heart muscle-fibres, many of which are large, have pronounced longitudinal striations. The intima of the small bloodvessels in the wall of the heart is thickened.

The condition of the lungs is that resulting from the inhalation of excessive amounts of some kind of dust accompanied by chronic passive congestion and slight emphysema. The most noticeable changes from normal are the masses of pigment and the thick bundles and irregular-shaped masses of connective tissue present. The pigment is found in the lymphatic spaces around the bloodvessels, and around the small bronchioles, and is irregularly distributed throughout the thick bundles of connective tissue. These connective-tissue masses contain thin walled bloodvessels and around the periphery of many of these masses are collections of lymphoid cells. In other places the alveolar walls are thickened; there has been a proliferation and desquamation of the epithelial cells lining these, so that the air-cells are more or less filled with them, and many contain pigment-granules, while in still other cells the alveolar walls have become thinned and broken, leaving emphysematous spaces. There were no tubercles found in the lungs.

The liver shows beginning cirrhosis, and in places slight chronic congestion.

Sections of the kidney show the organ to have been congested. There is an increase in connective tissue below the capsule and a slight increase between the uriniferous tubules in places, and the walls of a few of the glomeruli are thicker than normal. The intima of the small bloodvessels is thickened. In the tubules there is some granular detritus, but this condition is not marked, nor is the granular appearance of the protoplasm of the cells lining the tubules marked. One of the uriniferous tubules noticed in the pyramidal portion was distended, and contained a collection of red blood-corpuscles, epithelial cells and granular detritus. The epithelial lining of the distended tubules was in part intact, and in part had desquamated, leaving a fibrous wall to the small cyst thus forming. We have, therefore, a acute interstitis and a very slight parenchymatous degeneration.

Sections of the spleen show a thickening of the capsule and a slight increase of connective tissue throughout, with a diminution of spleen pulp. The blood-vessels have thickened walls and are filled with blood. The spleen, hence, shows chronic passive congestion.

In the testicle there are the absence of fibrous tissue between the tubules. In comparatively few of the tubules are forming spermatozoa seen, and none are found in the tubes of the epididymis, which is lined with cells larger than those usually found here, and which do not show cilia.

A transverse section was made of the posterior (tibial) artery and the tissues surrounding it (Plate II., Fig. 7). While the artery itself shows only a very slight thickening of the intima, the smaller vessels show a marked thickening, due to a cellular increase in the intima, the lumen below much narrowed and in some places almost occluded. The arrangement of the proliferated tissue making up this thickened intima is in papillary projections between which are narrow slits, as seen in Plate II., Fig. 7.

The skin of the great toe shows a thick epidermal layer, the dense intima of the small arteries and the connective tissue lined with a layer of epithelium several cells deep, instead of a single layer of cells as usual.

THE SKELETON. *The Skull* (Plate III., Figs. 8 and 9). The parietal and internal tables of the cranial bones are very thin, while the cancellous tissue of the occipital and frontal bones is much increased, making the bones as a whole somewhat thicker than usual. The middle fossa of the skull is very much deeper on the left than on the right side.

The supra-orbital ridges are much enlarged and prominent, while the glabella is very much enlarged, causing the forehead to appear very much to retreat, resembling the anthropoid apes. The metopic suture has entirely disappeared, as have also the sagittal and the lambdoid sutures. There are no Wormian bones present. The external occipital protuberance and the superior and inferior curved lines on the occipital bone are sharp and prominent. The vertical crest is very projecting and prominent. The mastoid processes are thin laterally, with sharp margins, the digastric fossa being very deep. The para-mastoid processes are not marked. All of the crests and eminences on the base of the skull are more developed and prominent than usual. The glenoid cavities are very deep, but narrow from before backward, the tympanic plate being very large. The styloid processes are somewhat elongated and completely ossified. The right external auditory meatus is considerably narrowed. The cells mastoid is greatly enlarged, having a cavity of one and one-eighth inches antero-posteriorly, one and one-half inches laterally, and seven-eighths of an inch perpendicularly.

The sphenoidal sinus of the left side is larger than the right, and further forward than usual, being pushed forward by the enlarged cavity of the cells

widened [illegible]. The
the enlargement of the mala

The measurements of th
being 44 mm. and the verti

The [illegible] canals are
wide at their upper orifices.

The posterior molar teeth
being [illegible] in position
facing outward, making a
process, while its [illegible] fang
between the tuberosity of th
mm. The inside measure
is 10 mm. The nasal proce
elongated, pushing the lowe
making a marked concavity
verse diameter of the anter
diameter is 40 mm. The a
ment with a very marked r
[illegible]. The incisor fossæ ar
much enlarged, the left bein

We therefore find the a

largement of the skull seems to be confined to the supra-maxillary and frontal bones.

Turning now to the inferior maxillary bone, we find the condyles to be elongated laterally and to project beyond the zygomatic processes. This enlargement seems to be due to the growth of the inner part of the condyles, which has thus pushed the rami outward. The transverse measurement of the left condyle is 57 mm. and the right 55 mm. The condyles are thin from before backward. The coronoid processes are not much enlarged, but are somewhat elongated. The angle of the jaw is quite obtuse. The rami are much lengthened, and the basilar border of the bone projects forward and is thick relatively to the rest of the bone. The mental region of this bone is greatly broadened. The width of the bone from angle to angle internally is 107 mm. (normal bone 88 mm.). From the outer edge of the one condyle to the outer edge of the other is 151 mm. (normal bone 120 mm.).

The Spinal Column (Plate IV., Fig. 10.) The spine is very typical of osteomalacy, showing a very marked kyphosis, the angle of which is in the upper dorsal region at the disk between the fifth and sixth dorsal vertebrae and is an angle of about 80 degrees. This angle of kyphosis was so great that when the body lay on the table at autopsy the head could not touch the table, but stood up. This enormous kyphosis seems to be due to an absorption and loss of the intervertebral disks anteriorly rather than to a slipping of the anterior portions of the bodies or a thickening of the posterior portions of the bodies of the vertebrae.

A scoliosis is present, but not marked, with its convexity to the right.

The second, third, fourth, fifth, sixth, seventh, eighth, ninth, and tenth dorsal vertebrae are connected by ossification of the anterior common ligament, there being marked exostoses at the position of the intervertebral disks. The second and third dorsal present anterior exostoses as wide as the bodies of the vertebrae. The second and third cervical vertebrae are ankylosed. The spinous processes are somewhat enlarged and show many projecting exostoses. The spinous processes of the fifth, sixth, seventh, eighth, and ninth dorsal vertebrae are partially ossified; the upper four apparently by an ossification of the supra-spinous ligament, while the eighth and ninth are joined by ossification of the interspinous ligament. The interspinous ligament of the seventh and twelfth dorsal vertebrae is ossified, binding these vertebrae together.

All of the lateral processes are roughened and thickened by increased cellular tissue. In fact, all of the spinal bones have less compact tissue than normal, especially the bodies, which are very spongy.

The whole larynx, which was examined with the tongue and epiglottis in position, was greatly enlarged. The width of the base of the tongue, dorsal surface, just in front of the anterior pillars was three inches. And here I might say that all of the curves of this region were enlarged. The greatest width laterally of the larynx was two and three-quarters inches.

The circumference of the cricoid cartilage was five and one-half inches.

The super ficial cartilages were also enlarged. From the superior border of the cricoid cartilage to the top of arytenoid was seven-eighths of an inch, including bone and soft tissue—i.e., vertical height of arytenoid. All of the cartilages show more or less calcification.

Below the vocal cords the larynx was much enlarged forming a widened space. The distance between the vocal cords being less than usual, owing to the enlargement of the soft tissues; so in spite of the larynx itself being enlarged.

The epiglottis is very prone.

The hyoid bone is horseshoe-shaped, the greater cornua looking toward each other rather than away from each other, as normal.

The hyoid bone from tip to tip around its anterior convexity measured seven inches.

The upper tracheal rings ossified.

The sternum is very greatly enlarged, measuring 230 mm. in length, with the greatest width of the gladiolus of 56 mm., and the width of the manubrium 78 mm. The angle between the manubrium and gladiolus is more projecting than usual. The manubrium and the first rib on each side are ossified. The sternum is thickened to its entire extent. The ensiform appendix is 65 mm. long, bone, and is completely ossified, and is consolidated with the sternum. There is a large foramen in the centre of the appendix.

The costal cartilages are ossified and are fully a third longer than usual. The sixth, seventh, and eighth costal cartilages are consolidated at intercostal joints on the left side, but are not consolidated on the right.

The ribs are broadened, thickened, and elongated with all curvatures lessened. The seventh rib measured 485 mm.

The clavicles (Plate V., Fig. 1) are enormously enlarged, being increased at least one-fourth in size. They are rather more curved than normal, with the margins of the articular surfaces prominent and extended. They present very large surfaces for articulation with the sternum.

The sternous extremity is very greatly enlarged.

The conoid, oblique, and trapezoid ridges are very much enlarged.

The scapulæ are markedly enlarged in every way except in length, and are distinctly broadened.

The glenoid cavities are somewhat deep and rough, with exostoses at the edges.

The spines are very large.

The heads of the humeri (Plate VI., Fig. 2) show roughened articular surfaces and an increase in bone, due to the ossification of the articular cartilages.

The deltoid ridges are enormously enlarged, evidently due to the ossification of the tendon of the deltoid. The same condition is true of the anterior oblique ridge at the site of the attachment of the pectoralis major muscle.

The external epicondylar ridge is very much enlarged and projecting.

The radii do not show roughened articular surfaces.

The interosseous ridge is very sharp and prominent. The lower ends of the radii are not enlarged, but have many bony protuberances.

The lower articular surfaces are widened by a growth of bone at the anterior margins.

The ulnar bones are enlarged at their upper ends, and are rough and irregular in all the details of their upper extremity. The articular surfaces are increased in extent by the growth of bone at their margins, especially at the coronoid process. This surface is also more concave and roughened than normal. The interosseous borders are enlarged. The lower extremity of the ulnar bone is flattened and not enlarged.

The articular surfaces encroach upon the normal depression between the head and styloid process. It is again an instance of the extending of articular surfaces by ossifications in cartilage.

The cavities of each are not enlarged, but the articular ends of the upper extremity of the ulna is enlarged.

All of the bones of the carpus (Plate VII., Fig. 13) are very much enlarged.

The metacarpal bones (Plate VIII., Fig 14), are all enlarged, especially at their upper extremities. The metacarpal bones of the thumbs are widened, not only at their extremities but in their shafts, due to a laying on of bone at the lateral ridges. The articular surfaces of these bones are also greatly enlarged.

The phalanges of the hand (proximal and middle) are not much enlarged, except that the ridges on these bones are more prominent than usual.

The distal phalanges show enlargement of both extremities, with a great deal of increase in spongy roughness at the ungual end. This end is broadened and thickened, due largely to ossification on the pulp surfaces.

The Pelvis. The whole of the iliac bones are enlarged.

The crests of the ilia are greatly thickened, and are not as sinuous as usual, while the iliac fossæ are more flattened than normal, apparently due to a spreading out of the ilia. The bones in the expanded portion of the ilia are thin. The distance from one anterior spine to the other is 395 mm.

The acetabular cavities are enlarged and very deep, with very prominent margins. The normal depression at the bottom of these cavities is three-quarters filled with an osseous plate, and the entire articular surface is rough.

The symphyses of the sacrum and ilia are partially consolidated anteriorly. The sacrum is broad and enlarged, due to an increase of the lateral masses. The concavity of the sacrum is somewhat increased. The coccyx and sacrum were not consolidated.

The pubic bones are thin, the obturator foramina being materially increased in size. The spines of the pubes are enlarged, especially on the left side.

The horizontal rami are thin and narrowed, and the bodies are thinner than normal. The distance from the symphysis anteriorly to the ilio-pubal eminence is increased, measuring 90 mm. The pubic bones have, therefore, lost in substance but have grown in extent.

In the left hip-joint in front anteriorly to the neck, just below the base of the femur, was found a bony irregular calcification one inch in diameter. The femurs are large with very large heads; the cartilages covering the heads are rough, and the articular surfaces are rough and porous. The great trochanters are very rough. The articular lamellæ of the condyles are rough and porous.

The tibiæ are not much enlarged; neither are the tibiæ, though the ridges of the tibiæ are prominent.

The patellæ are enlarged and show projecting processes and outgrowths from the upper and posterior borders. The margins of the articular surfaces show new bone-growth.

The os calcis is enlarged as a whole, but especially at its posterior surface at the attachment of the tendo-Achillis, shows an immense new ossification of a tendon. The whole bone is roughly irregular and nodular.

All of the bones of the tarsus are enlarged, showing enlargement of their articular surfaces by new laying on of bone, and showing roughening of their edges.

The sesamoid bones are not increased in number.

The metatarsal bones of the first and second phalanges of the toes are not enlarged, but the ridges are more marked and prominent than usual.

The third phalanges are enlarged at both extremities, with a growth of new bony bone not only at their distal ends, but at their proximal, with extension of lateral portions (see Plate V., Fig. 11) toward the distal ends, causing actual enclosed foramina on some of the bones, and deep notches in others, the majority, however, being closed foramina. The distal ends of these phalanges curve upward.

I am indebted to Dr. M. B. Ferris, Professor of Anatomy at Yale, for his careful supervision of the above detailed description of the changes from normal in the bones of this skeleton.

POINTS OF SPECIAL INTEREST IN THIS CASE.

The thyroid and the thoracic thyroid glands were both sent to Prof. R. H. Chittenden for chemical examination, accompanied by a positive statement that this man had received no iodine-carrying drug in at least six years.

Dr. L. B. Mendel, first assistant in Prof. Chittenden's laboratory, later sent me the following report, namely, that the thyroid gland contained but 0.15 of a milligramme of iodine as a maximum content for the total gland. This amount of iodine is, therefore, only one-twentieth of the amount which Baumann claimed the normal thyroid gland of the adult should contain—i.e., 3 milligrammes; while Weiss, from an analysis of fifty human thyroids, claims that the normal adult gland should contain 4 milligrammes.

Dr. Mendel also reports, in regard to the thoracic gland found in this case, that he not only found iodine in the gland, but in large amount, namely, 5 milligrammes as the content of the whole gland, or twice as much as is contained in the normal thyroid. He finds by careful analysis of different portions of this thoracic gland, that the whole of the iodine is contained in the cortical, or "more porous portion," while there is no iodine in the central, or "more compact and homogeneous" portion. This coincides exactly with the microscopical sections of this gland, which show the outer portion to be glandular and the central portion of the gland to be of connective tissue. (Plate II., Fig. 6.

So far as we are aware, this is the first time that iodine has been found in an auxiliary thyroid gland. It is needless to say that the very best possible control-tests were made. All of the preservative fluids and sections of other organs from the same jars, and the muscle-tissue around the glands, were all tested for iodine, with absolutely negative results.

We will turn now for a few minutes to the description of this case reported by me in 1892, in which, owing to the occasional small amounts of albumin being found in the urine "the kidneys were stated in all probability to be slightly damaged." The albumin was never found in any amount except on the day before death, and was absent, as were also casts, during the last few weeks of his sickness. Sugar was found once in small amount, about eight months before death. He complained of irregular abdominal pain, and the autopsy showed that there were a number of peritoneal adhesions.

The condition of the handle of the hammer of the right ear, discovered and remarked upon by Dr. H. L. Swain in 1892, was abundantly substantiated by the condition of the bone found at autopsy, namely, with its handle bent almost to a right angle.

In 1892 I wrote the following in regard to the pituitary body of this patient:

"Have we in this case the symptoms of an enlarged pituitary body? Yes; the enlarging gland would naturally tend to press in the direction of the least resistance. The least resistance can only be out of the sella turcica, either laterally toward the cavernous grooves, or upward and forward toward the middle clinoid processes. The excess of pressure in either direction will depend upon the bony environments of the sella turcica, and in either case the enlarging pituitary body reaches bloodvessels. If the bone formation of the sella turcica is such that the first pressure be exerted laterally, the internal carotids and cavernous sinuses are pressed upon and the soft walled sinus is most affected. By this pressure we then have a venous hum in the ears, and the pressure being continuous the tinnitus will be continuous. The difference in the intensity of the ringing in the ears might be due to an asymmetrical enlargement of the pituitary gland, but is more probably due to the formation of the sella turcica allowing freer passage of the gland to one side than on the other. The middle clinoid process being more prominent on the one side than on the other a frequent condition—the enlarged gland would press more on the opposite bloodvessels. This is probably the condition in this case."

The above inference has certainly been well borne out by the conditions found at autopsy. A small-celled tumor-growth containing a cyst occupied the sella turcica; this could not press upward owing to the bony formation of this sella turcica, but did grow laterally and much more on the right side. This tumor-growth not only pressed upon the right cavernous sinus, but also this small-celled growth had grown into the wall of the sinus, almost occluding it. The same was true, but in much less degree, of the left cavernous sinus. This then gives the cause of the constant venous buzzing or humming of the ears, which from the loss of gravity in these veins would be worse on lying down.

In 1888 this man showed decided symptoms of brain pressure, and at that time the diagnosis of "brain tumor" was made; the acute symptoms (however, lasting only about one month) I stated in 1892 to have been probably "due to the first pressure of the enlarging pituitary body on the bloodvessels," which gave him "continuous head pain and caused him to have deranged spells."

In 1892 I described the pain, which he said he constantly had and referred to the top of the head at the region of the anterior fontanelle which is directly perpendicular to the sella turcica, and about the size of the thumb tip, and also referred down the right side of his nose. I attributed this to the pressure of the enlarged pituitary body. It is interesting now to note that directly under this spot was a plate of ossification in the dura mater.

This man died by syncope, and unexpectedly, as Marie claimed to be typical of death from acromegaly. This was evidently due, as was the edema of the feet and legs, to his weak heart. The heart con-

tained connective tissue between its fibres, and some fatty globules;
hence, though the heart was enormous in bulk it was weak in action,
as was also shown by the auscultatory sounds.

The interesting factors in this case might be summed up as follows:

1. The liver weighed seven pounds, two ounces; a normal liver
three to four pounds.

2. The spleen weighed two pounds, four ounces; a normal spleen
weighs seven ounces.

3. The heart weighed two pounds, nine ounces, one of the largest
hearts on record; a normal heart weighs from ten to twelve ounces.

4. A thoracic thyroid gland weighing thirty-six and one-half
grammes, containing a large amount of iodine.

5. The thyroid gland weighed over 101 grammes, with almost
absence of iodine; the normal thyroid weighs from thirty to sixty
grammes.

6. General increase of intima of all bloodvessels.

7. Increase of connective tissue in the heart.

8. Small-celled growth in sella turcica, and encroaching upon cavernous sinus, and same growth in small tumors attached to base of
brain (sarcomatous?).

9. Sweat-glands lined with extra layer of epithelium.

10. Large sella turcica.

11. Large superior and inferior maxillary bones.

12. Clavicles enormous.

13. Spine typical of acromegaly.

14. Carpus enlarged.

15. Tarsus enlarged.

16. End-phalanges of feet show peculiar bone-growth.

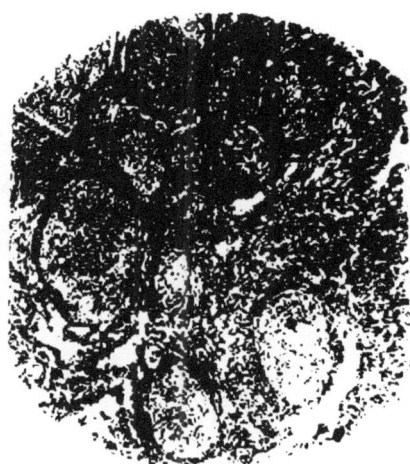

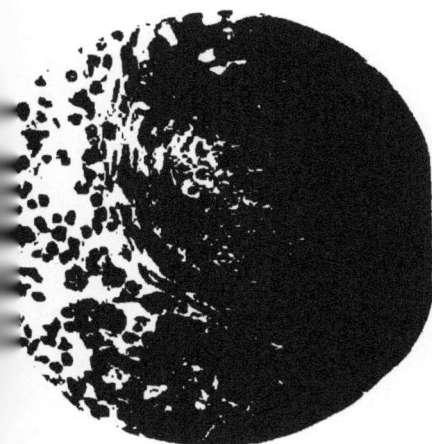

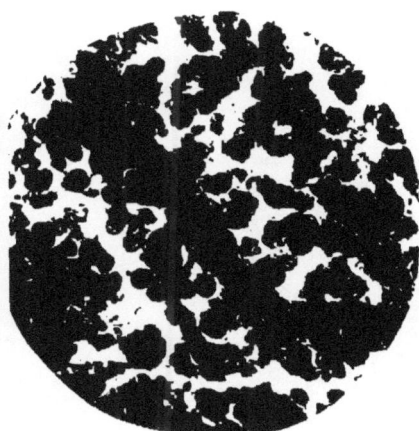

Fig. 1.

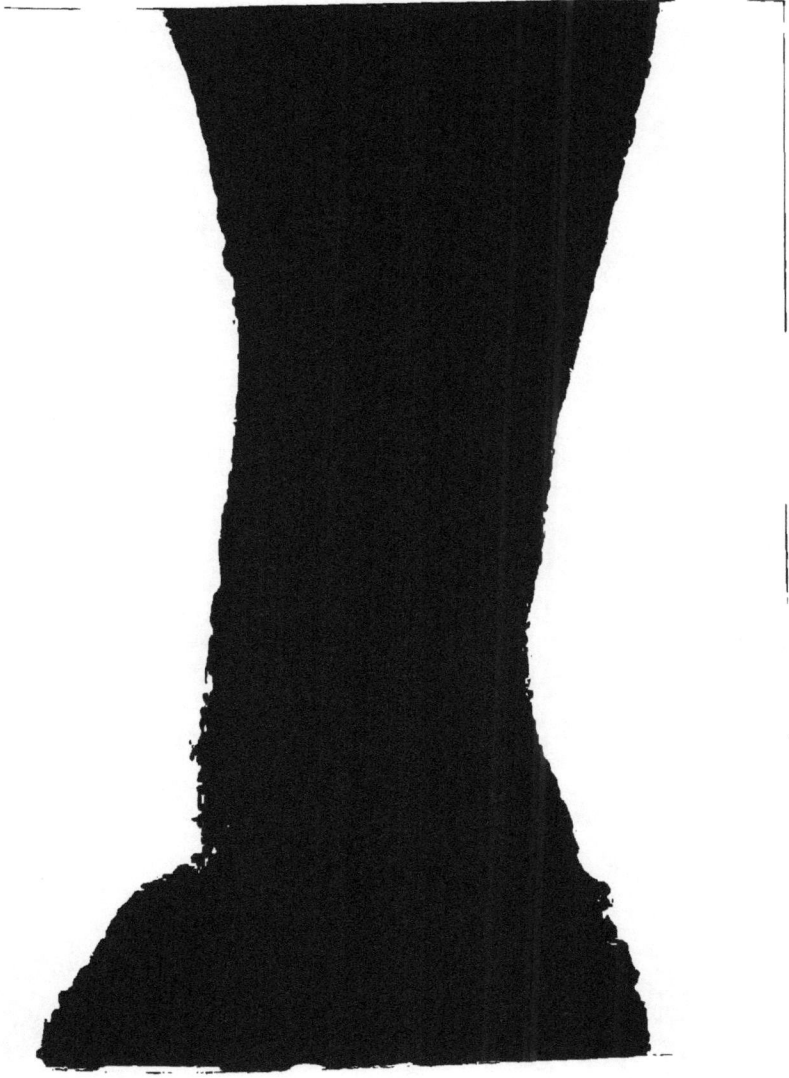

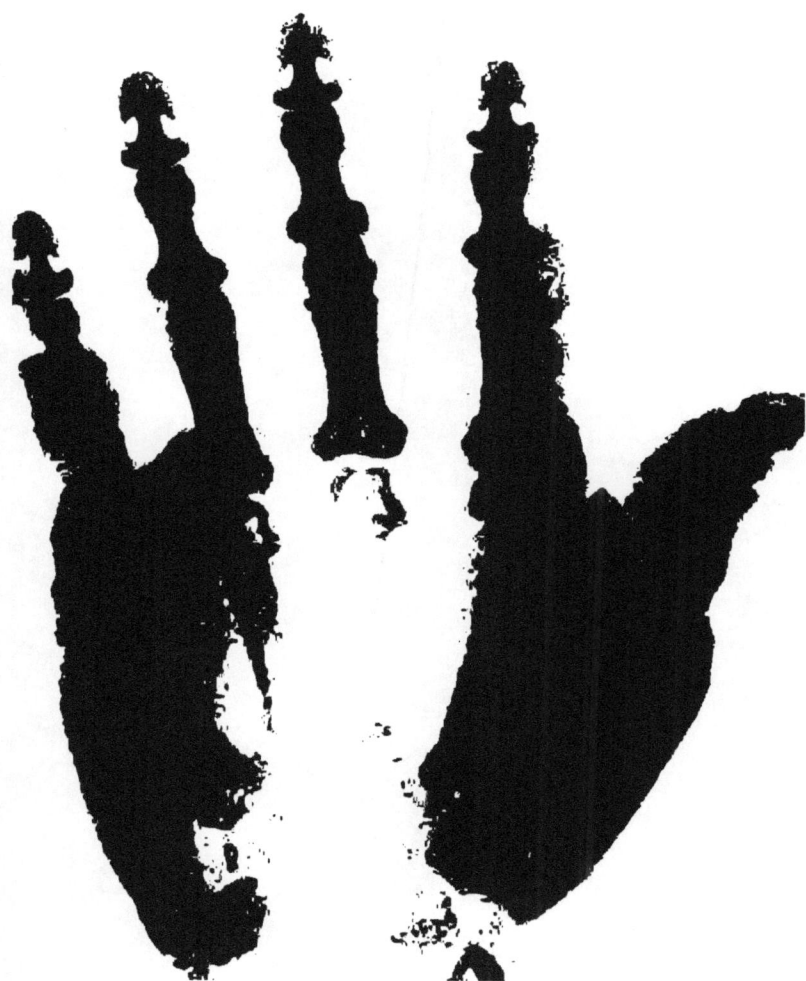

www.ingramcontent.com/pod-product-compliance
Lightning Source LLC
Chambersburg PA
CBHW022033190326
41519CB00010B/1695